W9-BGP-229

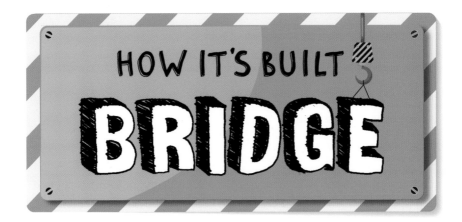

HOW IT'S BUILT
BRIDGE

by Vicky Franchino

Illustrations by Richard Watson

Children's Press®
An imprint of Scholastic Inc.

Thanks to Reed Brockman, Associate VP/Senior Structural Engineer, New England Transportation, for his role as content consultant for this book.

Thanks to Donna Lowich, Senior Information Specialist at the Christopher & Dana Reeve Foundation, for her insights into the daily lives of people who use wheelchairs.

Thanks to the New York State Thruway Authority for their support and expertise. Although this is a work of fiction, the majority of photos, including the cover, are from construction of the Governor Mario M. Cuomo Bridge in New York.

Library of Congress Cataloging-in-Publication Data
Names: Franchino, Vicky, author. | Watson, Richard, 1980– illustrator.
Title: How it's built: bridge/by Vicky Franchino; illustrated by Richard Watson.
Description: First edition. | New York: Children's Press, an imprint of Scholastic Inc., 2022. | Series: How it's built | Includes index. | Audience: Ages 5–7. | Audience: Grades K–1. | Summary: "Narrative nonfiction with fictional characters who visit various work sites to find out how different structures are built. Full-color illustrations and photographs throughout"—Provided by publisher.
Identifiers: LCCN 2021029565 (print) | LCCN 2021029566 (ebook) | ISBN 9781338800111 (library binding) | ISBN 9781338800128 (paperback) | ISBN 9781338800135 (ebk)
Subjects: LCSH: Bridges—Design and construction—Juvenile literature. | BISAC: JUVENILE NONFICTION / Technology / How Things Work—Are Made
Classification: LCC TG300 .F74 2022 (print) | LCC TG300 (ebook) | DDC 624.2/5—dc23
LC record available at https://lccn.loc.gov/2021029565
LC ebook record available at https://lccn.loc.gov/2021029566

10 9 8 7 6 5 4 3 2 1   22 23 24 25 26

Printed in the U.S.A.   113
First edition, 2022

Series produced by Spooky Cheetah Press
Book design by Maria Bergós, Book & Look
Page design by Kathleen Petelinsek, The Design Lab

Photos ©: cover, 6–7: New York State Thruway Authority; 8 right: Anastasia Yakovleva/Dreamstime; 9 bottom: Krzysztof Nahlik/Dreamstime; 10–11: fxoleary/Getty Images; 10 inset: Imaginechina Limited/Alamy Images; 12–13 bridge: New York State Thruway Authority; 14 left: Keren Su/China Span/Alamy Images; 15 bottom: New York State Thruway Authority; 15 top left: Serhii Chrucky/Alamy Images; 16 right, 17 left, 18–19: New York State Thruway Authority; 19 inset: Galen Rowell/Mountain Light/Alamy Images; 20–28 all, 29 top left, 29 bottom left, 29 bottom right: New York State Thruway Authority; 30 top left: David R. Frazier Photolibrary, Inc./Alamy Images; 31 bottom left: robertharding/Alamy Images.

All other photos © Shutterstock.

TABLE OF CONTENTS

MEET THE JUNIOR ENGINEERS CLUB

Sofia
Lucas
Kai
Nisha
Jacob
Zoe

These six friends love learning about how things are built! This is their workshop.

Zoe and Kai found out how a bridge is built. Now they are sharing what they learned!

Even a small bridge takes months to design and build. A large bridge can take many years.

PROJECTS
HOUSE
CAR
BRIDGE
SKYSCRAPER
ROCKET
SAILBOAT

• LET'S BUILD A BRIDGE! •

Hi! I'm Zoe, and this is my friend Kai. Our club went to the grand opening of the new cable-stayed bridge in our city. But first, Kai and I wanted to know how bridges are built. We met Johanna, an architect, and Layla, an engineer. They told us that all bridges have a few things in common.

Old bridge

New bridge

Sometimes a new bridge will replace an old bridge. The old bridge will not be torn down until the new bridge is ready. That way, people won't have to take a different route to get where they are going.

The **deck** is the part that people walk or drive on. A deck that people drive on is called a roadway.

If bridges are built for vehicles, they need to have **roadway lights** to help drivers see at night.

If bridges are high, they need to have **extra lights** strung along their tops. This keeps planes from hitting them!

A **tower** is a special pier that extends far above the road.

A **pier** is any support that holds up the bridge. Typically, piers are found beneath the deck or roadway.

An **abutment** is found at each end of a bridge. It holds up the end of the bridge and keeps the deck from crumbling into the water.

Johanna and Layla were part of the large team that worked on the new cable-stayed bridge. They told us that there are several types of bridges.

Abutment

A **beam bridge** is the simplest type. A short beam bridge has a support, or abutment, only at each end. A longer beam bridge has piers in more places in between the abutments.

An **arch bridge** looks like the top half of a circle. Each end of the curve is anchored to an abutment.

Some bridges are built over land, and some are built over water.

Our new bridge was built over a river.

A **truss bridge** is a special type of beam bridge. The top and bottoms of the beam are pinned together with more beams. A truss bridge can be longer than an arch bridge or a simple beam bridge.

A **suspension bridge** gets its name because the deck hangs from a pair of suspension cables that run over the towers. Shorter vertical cables, called suspenders, are spaced evenly along the length of the bridge. They tie the deck to the main suspension cables.

A **cable-stayed bridge** has towers just like a suspension bridge does, but it doesn't have hangers. The deck is tied directly to the towers by a series of long cables called stay cables.

There's a lot to think about when designing a bridge. Form is what the bridge will look like. The cable-stayed design of our bridge is very beautiful. Function is how the bridge will be used. A lot of cars and trucks traveled over the old bridge every day. The same will be true for the new bridge, so the roadway has to be very wide.

Some bridges, like the Lucky Knot Bridge in China (above), are built just for people to walk on. They do not need to be as strong or as wide as bridges that cars will drive on.

Old bridge
My mom drove over the old bridge on her way to work every day.
We crossed it when we visited my grandparents on the weekends.
And soon everybody will be able to use the new bridge!

It is the work of engineers like Layla to make sure that the bridge is strong. It must be strong enough to hold itself up. It also has to support the people and vehicles that use it.

8 towers anchor the stay cables

192 stay cables support the 74-million-pound deck

82 piers support the bridge from below

A force is something invisible that can push against or pull on a bridge. Wind is one of those forces. Engineers design a bridge so that it can move a little bit, but not too much!

This is a 3D drawing, or rendering, of our new bridge.

Wow! It's so detailed!

13

Layla and Johanna told us that most bridges are made from steel or concrete. Those are what the team chose for the new bridge. But bridges can be made out of many different materials.

Rope bridges move a lot as people cross them. The Q'eswachaka rope bridge (pictured) in Peru is more than 500 years old. It is the last remaining Inca bridge! Every year people from the surrounding communities get together to repair the bridge.

Wood can be strong, and people like how wooden bridges look. But wood is not as strong as concrete, iron, or steel. Wooden bridges can be destroyed by bad weather and bugs. Ponte degli Alpini (pictured) is a wooden bridge in Italy.

Steel is used for different parts of a bridge. It is very strong and can easily be made into many shapes. Eads Bridge in St. Louis, Missouri (pictured), was the first steel bridge.

Iron is a very strong material, but it is not as strong as steel. Iron is not as easy to work with, either. The Iron Bridge in England (pictured) was the first bridge made of iron.

I would *not* want to drive a car over a rope bridge!

Concrete is the most common material used to build bridges. Reinforced concrete is used for long spans. Steel is added to the concrete to make it stronger. The Governor Mario M. Cuomo Bridge in New York (pictured) was built from reinforced concrete and steel.

After the bridge was designed, it was time for the general contractor, Rainey, to take over. He worked with all the people who built the bridge. The first step was to build strong piers to hold up the bridge. Building a bridge over water is a tricky job!

1

First, workers sank a caisson. That is like a box without a bottom. The caisson was pushed down into the water. The water in it was removed and air was trapped inside. Workers could go inside the box to work and stay dry!

2

Then workers drove piles deep into the ground. Piles are long poles. For this bridge, the piles were made of steel.

3

Then a pier was set on top of the piles. The pier holds pieces of the bridge. The piers on our new bridge were made of reinforced concrete.

Our bridge is very long, so it needs a lot of piers.

After the piers were built, workers laid long beams of steel on top. These beams are called girders. Then the bridge deck and towers were built. Both are made of reinforced concrete. Every part of the bridge needs to be strong!

The workers look so small compared with the bridge!

It's a *really* long bridge!

Workers

Bridges can be damaged by wind, sun, rain, and snow. Painting a bridge helps protect it. The Golden Gate is a famous bridge in California. It is such a big bridge that workers have to touch up the paint on some part of the bridge every day!

There are different ways to attach the cables to the tower and the deck. On this bridge, all the cables attach near the same point at the top of the tower.

Deck

Next, all the cables were added. They are made of strands of steel twisted together. The cables attach the deck to the towers.

Then workers added railings to the bridge. Railings are a safety feature. They keep people and vehicles from falling off the bridge.

Lots of people worked together to build the new bridge. All the expert workers had special training to learn how to do their jobs.

Welders used electricity to produce heat to join metal parts.

Crane operators ran cranes, the big pieces of equipment that moved large objects at the building site.

Ironworkers connected the parts of the bridge that were made of steel. A long time ago, many parts of bridges were made of iron. Now steel is used instead of iron, but the name "ironworker" is still used!

Concrete finishers poured wet concrete and made it into different shapes.

Electrical workers installed and fixed the bridge's wiring and lights. They made sure the bridge has electrical power.

We used a large, flat boat called a barge.

How did you transport materials over the water?

Riggers brought heavy equipment and materials wherever they were needed.

Once the new bridge was complete, it was time to demolish the old bridge. People with special training used explosives to blow it up. They had to be very careful! They worked hard to protect the people who were nearby and the fish that lived in the river. They also had to protect the new bridge.

Old bridge

New bridge

Chains were used to pull pieces of the old bridge out of the water. Sometimes materials from an old bridge can be reused or turned into something else.

On the day of the grand opening, all the members of the junior engineers club met at the bridge. Many people came to see the new bridge and walk or drive across it. We went across, too! Knowing just how this bridge was built made crossing it extra exciting.

MACHINERY AND TOOLS FOR BUILDING A BRIDGE

Dredging Buckets
These machines clean out mud, weeds, rocks, and garbage in the water where the bridge will be built.

Excavators
These heavy machines help tear down the old bridge when it is demolished.

Tugboats and Barges
Small tugboats (pictured) help guide the larger barges that bring heavy equipment to the bridge.

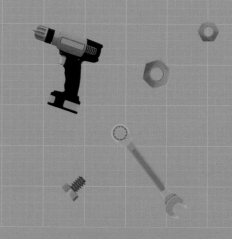

Cranes
These huge machines raise and lower materials at the bridge site.

Generators
Most tools need power to work. Generators make power from gasoline. Workers can plug their tools into the generators for a recharge.

Concrete Mixers
When a bridge is built over water, a barge carries the concrete. The barge has a long arm with a hose so the concrete can be poured high above.

Safety Gear
Hard hats protect workers' heads from falling materials. Safety harnesses secure the workers to the structure they are building.

BRIDGES BUILT IN AMAZING WAYS

The Lake Pontchartrain Causeway in Louisiana is a beam bridge that is 24 miles (39 kilometers) long. It is the longest bridge in the United States. The bridge was built in pieces on land and carried by boat.

The **Golden Gate Bridge** is painted red-orange so it is easy to see in the fog. Its giant towers were built underwater. Divers had to go deep in the water to find good places where the future bridge could be attached.

The **Millau Viaduct** is a cable-stayed bridge in France. It is the tallest bridge in the world. The deck is 886 feet (270 meters) above a river. Workers built the deck pieces in a different city. They brought them to the building site on trucks.

The **Sydney Harbor Bridge** in Australia is the world's largest steel arch bridge. This was a tricky bridge to build because it had to go over a wide body of water. Different engineers thought about how to build this bridge for more than 100 years! Finally, people learned how to make steel and reinforced concrete. Those were the materials used when building began on this bridge in 1923.

The **Akashi-Kaikyo Bridge** is in Japan. At 6,532 ft. (1,991 m) long, it is the longest suspension bridge in the world. Workers used giant caissons during building.

Ponte dei Quattro is made of stone and rock. It is the oldest bridge in Rome, Italy, that is still being used. It was built in 62 BCE.

Tower Bridge is a drawbridge in London, England. It was built between 1886 and 1894. Drawbridges used to be popular because they helped protect people. People could pull up the drawbridge to stop others from coming into their city.

INDEX

ABOUT THE AUTHOR

Vicky Franchino has written many books for children. The most unusual bridge she has ever walked on is the Carrick-a-Rede in Northern Ireland. It is made of rope. Vicky was very brave and walked across it even though she was a little scared!